Chemical Pesticide Markets, Health Risks and Residues

CABI *Bioscience* is a division of **CAB** *International*, an inter-governmental, not-for-profit, mission-oriented organization dedicated to improving human welfare world-wide through the dissemination, application and generation of scientific knowledge in support of sustainable development. Emphasis is placed on agriculture, forestry, human health and the management of natural resources and particular attention is given to the needs of developing countries.

CABI *Bioscience*'s **Biopesticides Programme** is committed to the development and use of biopesticides as safe, environmentally friendly alternatives to chemical pesticides. The Programme carries out collaborative inter-disciplinary research and development, offers training in insect pathology, runs the International Biopesticide Consortium for Development (IBCD), disseminates information and promotes the role and value of biopesticides in sustainable crop production, poverty alleviation and wealth generation.

Biopesticides series

1. *Chemical Pesticide Markets, Health Risks and Residues*
   J. Harris

2. *Priorities in Biopesticide Research and Development in Developing Countries*
   J. Harris and D.R. Dent

Biopesticides are biological pesticides based on beneficial insect and weed pathogens and entomopathogenic nematodes. Pathogens used as biopesticides include fungi, bacteria, viruses and protozoa. Produced, formulated and applied in appropriate ways, such biopesticides can provide ecological and effective solutions to pest problems.

The aims of the Biopesticides series are to more widely appraise and promote the role and value of biopesticides as alternatives to chemical pesticides and to improve awareness of the opportunities offered by biopesticides.

The series has been developed by the Biopesticides Programme at CABI *Bioscience* as part of its mission to disseminate information and promote the role and value of biopesticides.

Chemical Pesticide Markets, Health Risks and Residues

Biopesticides Series No. 1

Jeremy Harris

CABI Bioscience (UK Centre)
Ascot, UK

CABI *Publishing*

CABI *Publishing* **is a division of CAB** *International*

CABI Publishing
CAB International
Wallingford
Oxon OX10 8DE
UK

Tel: + 44 (0)1491 832111
Fax: +44 (0)1491 833508
Email: cabi@cabi.org
Web site: http://www.cabi.org

CABI Publishing
10 E 40th Street
Suite 3203
New York, NY 10016
USA

Tel: + 1 212 481 4018
Fax: + 1 212 686 7993
Email: cabi-nao@cabi.org

A catalogue record for this book is available from the British Library, London, UK.

Library of Congress Cataloging-in-Publication Data

Harris, Jeremy.
Chemical pesticide markets, health risks and residues/Jeremy Harris.
p.cm.
Includes bibliographic references and index.
ISBN 0-85199-476-8 (alk. Paper)
1. Pesticides--Toxicology, 2. Pesticides--Developing countries. 3. Pesticide residues--Developing countries. 4. Pesticides--Environmental aspects. I. Title

RA1270.P4 H365 2000
363.738'4--dc21
00-031184

ISBN 0 85199 476 8

Printed and bound in the UK by Cromwell Press, Trowbridge, from copy supplied by the author.

Contents

Preface..vii

Introduction..1

Chemical Pesticide Markets...3
    Global figures..4
    Exports from developed countries..5
    Markets: Latin America...6
    Markets: Asia...10
    Markets: Africa..13

Pesticide Exposure and Health Effects..15
    Latin America...16
    Middle East..23
    Asia..23
    Africa...31
    Europe..34
    North America..34

Pesticide Residues..37
    Latin America...38
    Asia..38
    Africa...40
    Europe..41
    North America..41

Obsolete Pesticide Stocks...45
    Africa...47

Asia 48

Index 51

Preface

The information available on the problems associated with the use of chemical pesticides in developing countries is limited and widely distributed amongst different sources, although through the efforts of the two newsletters, *Pesticides News* and the *Global Pesticide Campaigner*, in publicising these issues, this type of information has become easier to find over the last ten years. As part of CAB *International*'s mission to provide information which is relevant to the needs of developing countries, it was felt that there was a need to collate in one document the existing information concerning these issues in order to give an indication of the scale of the problems which exist for developing countries.

This bibliography provides information on the size and trends of the pesticide markets, relates cases of pesticide poisoning through occupational exposure and food residues, and describes the problems with storing obsolete chemical pesticides in developing countries. It is by no means complete. Developing countries have few resources to monitor these issues and therefore the information that could be found was often limited to one-off studies and anecdotal accounts. It does however provide a small picture of the significant problems that exist for developing countries and it is hoped that this document will be updated as new information arises.

Introduction

There has been a substantial increase since the 1940s in chemical pesticide sales and use in both developed and developing countries and predictions are that pesticide use will continue to grow over the next five years. The increasing use of a wide range of toxic chemicals deliberately released into the environment is causing widespread concern about their impact on human health and the damage caused to the environment (Table 1), particularly in developing countries which usually lack appropriate resources to minimise the risks and rectify problems.

The purpose of this bibliography is to provide a collation of:

- information on the scale of manufacture, import, export and use of chemical pesticides in developing countries;
- examples of the direct risks to human welfare frequently observed in terms of acute poisonings caused by occupational exposure and consumption of pesticide residues in food; and
- examples of problems with the storage of obsolete stocks of pesticides in developing countries.

The information presented indicates that the problems of chemical pesticide poisoning and storage are associated with significant health care and clean up costs which developing countries can rarely afford. In addition, the lack of resources to monitor the situation effectively means that the real extent of the problems is unknown. Some accounts have been included from developed countries to show that even when the resources such as training, equipment and effective regulation of pesticides are available, similar acute problems can still occur despite efforts to minimise the risks. The implication is that the problems will occur much more frequently and on a larger scale in developing countries where cheaper but more hazardous pesticides are regularly applied with substandard equipment.

The bibliography is not intended to be exhaustive. Other important problems such as chronic health effects, environmental persistence, bioaccumulation and pest resistance also exist. However, these are not included here as they are considered to be chronic problems equally applicable to both developed as well as developing countries.

Table 1. Some pesticide statistics.

☠	Despite a ten-fold increase in the use of chemical insecticides since WW2, the loss of food and fibre crops to insects has risen from 7% to 13%.
☠	In 1985, the WHO estimated that there are 3 million acute, severe pesticide poisonings and 20,000 accidental deaths each year. In 1990, the WHO revised their estimates to 25 million cases of acute occupational pesticide poisoning in developing countries each year[1].
☠	An International Labour Organization report of 1996 draws attention to dangers in the agricultural sector, where 14% of all known occupational injuries and 10% of all fatal injuries are caused by pesticides.
☠	60 pesticide active ingredients have been classified by recognised authorities as being carcinogenic to some degree. 118 pesticides have been identified as disrupting hormonal balance.
☠	In 1996, ten companies controlled over 80% of the global agrochemical market, valued in 1995 at US$30 billion. 25% of agrochemical sales are in developing countries and this is increasing.
☠	The quantity of obsolete pesticides in Africa alone is more than 20,000 tonnes, which will cost up to US$150 million to destroy.

Source: *The Pesticides Trust Review 1996*, except [1]Jeyaratnam, (1990) *World Health Statistics Quarterly* No. 43.

Chemical Pesticide Markets

The information presented in this section clearly shows an increase in pesticide sales and use globally and in individual developing countries during the last two decades. This increase is expected to continue into the next decade as more farmers adopt "conservation tillage" practices and genetically modified crops which allow increased applications of pesticides.

Large quantities of pesticides are exported from developed countries to developing countries and these exports are increasing each year. Many of these exported pesticides have either been banned, restricted or not registered for use in the country of manufacture. Exports include significant amounts of chemicals classified as "hazardous" or "extremely hazardous" to human health by the World Health Organization (WHO). Some developing countries such as India have the capability to manufacture their own pesticides and use or export these pesticides, many of which have been long banned in developed countries.

Pesticides in developing countries are sprayed on a variety of food (e.g. rice, maize, soybean) and non-food crops (e.g. cotton and tobacco) of which large amounts are exported to developed countries. It appears that a significant proportion of the pesticides used in developing countries on these crops are WHO class Ia (extremely hazardous) or class Ib (highly hazardous). Although the Food and Agriculture Organization of the United Nations (FAO) recommends that WHO Ia and Ib pesticides should not be used in developing countries, they are frequently cheaper than less hazardous alternatives and therefore they are often used.

Global figures

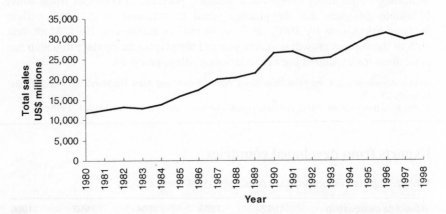

Figure 1. Global pesticide sales 1980-1995 (source: British Agrochemical Association annual reports 1980–1998. (Adapted from data tables in *Pesticides News* 28, June 1995 and from Bateman, R. (1999) *Rational pesticide use: targeting a way of the treadmill. In*: Proceedings of Conference on Managing Risks in the Use of Chemicals for Agriculture and Public Health, Agriculture Institute of Malaysia, Kuala Lumpur, pp. 31–36).)

It has been predicted that the global agrochemical market will grow by 1.9% a year between 1995 and 2005 to US$36.8 billion.

Agrow, *World Crop Protection News*, No. 276, March 14, 1997

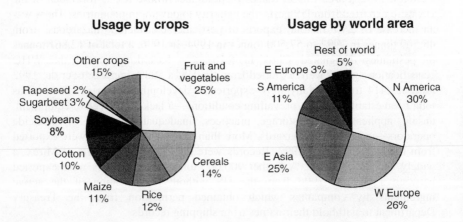

Figure 2. World pesticide usage (source: *Pesticides News* 32, June 1996).

The global agrochemical sales of the top ten agrochemical companies all showed an increase in 1996 and 1997. The planting of genetically modified herbicide tolerant crops offers potential for continued expansion in sales of herbicides although most of the increase in the next few years is expected to come from increasing adoption of "conservation tillage" practices in countries world-wide. Monsanto estimates that the practice could be extended to up to 97 million hectares world-wide by 2000, up from 74 million hectares in 1997. More than 40% of the volume growth in recent years of the glyphosate product Roundup has come from the expanded use of conservation tillage practices.

Source: Agrow, *World Crop Protection News*, April 18 1997,and April 10, March 27, March 13 and February 27 1998
Pesticide Action Network North America Updates Service, April 30, 1997 and May 6, 1998

Exports from developed countries

Table 2. Reported exports from US ports, 1992–1996 (kilograms).

Pesticide category in the US	1992	1993	1994	1995	1996
Banned	2,666,962	2,205,659	3,840,938	2,923,298	2,312,678
Discontinued/severe restriction	2,768,623	3,744,737	2,068,932	2,641,457	2,620,449
Never registered	2,043,857	1,113,556	1,338,447	2,184,423	2,041,659
Restricted use	25,993,189	31,807,412	32,482,937	33,249,120	35,595,153
Total	33,472,631	38,871,364	39,731,253	40,998,299	42,569,939

Source: *Pesticides News* 32, June 1996 and *Pesticides News* 40, June 1998

Between 1992–1996, 13,950 tonnes of pesticides whose use is forbidden in the US were exported from US ports, the majority to developing countries. There was an increase of 26% in total exports of pesticides considered "hazardous" from 45,500 tonnes in 1992 to 57,500 tonnes in 1994. In 1996, a total of 12,861 tonnes of pesticides designated Class Ia (extremely hazardous) under the WHO classification system were exported from the US, a 500% increase over the 1992 total of 2414 tonnes. Many were exported to developing countries where it has long been established that prevailing conditions—a lack of protective equipment, unsafe application and storage practices, inadequate training of pesticide operators—increase their hazards. More than 270 tonnes of DDT were exported from the US to Peru in 1992. Records were also found of exports of Mirex, a widely banned pesticide. A further 900,000 tonnes of pesticides were exported between 1992 and 1996 from the US without identification of the active ingredients by companies which obtained permission from the Treasury Department to withhold their names from shipping records.

"Exporting risk: pesticide exports from US ports, 1992–1994", Foundation for Advancements in Science and Education (FASE), Research Report, Spring 1995.
Pesticides News 32, June 1996, Page 16.
"Exporting risk—US hazardous trade 1995–1996", *Pesticides News* 40, June 1998, page 4

For many years, the UK has exported more pesticides than it uses with 67% of its total sales of £1431 million in 1995 destined for overseas markets. 1995 saw a real increase of 5.5% in total pesticide sales on 1994, with exports rising by £94.8 million (10.9%) to a record £967.6 million.

British Agrochemicals Association (BAA), Annual Review and Handbook, 1996.
Pesticides News 32, June 1996, Page 16

In 1991, Japan exported a total of 50,000 tonnes world-wide, including almost 2000 tonnes of pesticides to Africa worth US$27 million and over 2100 tonnes to Latin America worth US$40.9 million. Of the total, about 8% was exported as aid, mostly to Africa; of the 29 countries receiving aid in 1991, two were in Asia and 27 in Africa. Japanese aid has allowed chemical pesticide producers to gain a substantial market share in some developing countries and has accounted for 90% of the Japanese exports to Africa.

Challenging Japan's Pesticide Aid, *Global Pesticide Campaigner* 1993, Vol.3 No.2

Markets: Latin America

Argentina

Pesticide sales in Argentina were US$521.5 million in 1994, up 50% from 1992. More than half the agrochemical inputs are for soybeans which comprise the largest area of Argentina's crops.

"Crop Protection in Latin America", Agrow Reports, 1996
Pesticide Action Network North America Updates Service, April 16, 1996

In 1996, approximately 120,000 hectares of herbicide tolerant soybeans were planted in Argentina. These crops offer potential for the increased sales and use of herbicides.

Agrow, *World Crop Protection News*, No. 278, April 18, 1997

Aldicarb (WHO Class Ia) is banned in Argentina but exports from the US have been noted of over six US tons per month, a total of more than 300,000 pounds in weight in both 1995 and 1996.

Dirty Dozen Pesticides: Banned but Still Traded, *Global Pesticide Campaigner* 1999, Vol. 9 No. 1

Brazil

Pesticide sales in Brazil were predicted to increase by around 15% in 1998 to US$2153 million based on provisional figures from the country's industry association, ANDEF. Herbicides account for 60–65% of agrochemical sales, followed by insecticides at 25%. Insecticide sales have shown the greatest growth rate with a 19.8% increase over 1997.

Agrow, *World Crop Protection News*, No. 318, December 11, 1998

Brazil accounts for 55% of pesticide sales in Central and South America. Pesticide sales in Brazil were US$1.4 billion in 1994 and an increase in sales was seen for all leading crops between 1993 and 1994 (up 57% for cotton, 44.8% for coffee, 36.3% for maize, 30.2% for soybean, 17.7% for citrus fruits and 11.5% for sugarcane). Herbicide sales dominate the market (50% in 1994) and it was predicted that usage would continue to expand as more farms convert to minimum tillage practices (due to severe problems with soil erosion and deterioration in soil structure) that rely on increased herbicide applications to control weeds.

"Crop Protection in Latin America", Agrow Reports, 1996
Pesticide Action Network North America Updates Service, April 16, 1996

Chlordane and heptachlor are banned in Brazil but exports from the US have totalled 2 million pounds and 129,900 pounds respectively during 1995–1996.

Dirty Dozen Pesticides: Banned but Still Traded, *Global Pesticide Campaigner* 1999, Vol. 9 No. 1

Chile

Pesticide imports into Chile more than doubled between 1984 and 1996 from 5500 to 13,000 tonnes.

Market Opportunity Brief, Joint Environmental Markets Unit (JEMU), Chile, Technology Partnership Initiative (TPI), April 1996
Pesticides News 37, September 1997, Page 8

Colombia

25,423 tonnes of pesticide active ingredients were produced and 20,642 tonnes sold in Colombia in 1989, up from the 18,154 tonnes produced and 17,853 tonnes sold in 1985, according to government data.

"Profile: Pesticides in Colombia", *Global Pesticide Campaigner*, Vol.1 No.3

Pesticide sales in Colombia were US$316.2 in 1994, up 19.8% from 1993.

"Crop Protection in Latin America", Agrow Reports, 1996
Pesticide Action Network North America Updates Service, April 16, 1996

Costa Rica

The value of Costa Rica's chemical pesticide imports in nominal terms increased from US$56.2 million in 1990 to US$84.2 million in 1994 (almost a 50% increase). Costa Rica is a growing market for chemical pesticides, above all for fungicides whose imports almost tripled from US$14.9 million in 1990 to US$42.5 million in 1994.

The total volume of pesticide imports increased from 10.3 million units (kilograms and litres) in 1990 to 25.3 million units in 1994, an increase of 146%, although this increase was mainly caused by expanding imports of agricultural mineral oils which were classified as adjuvants.

Imported quantities of fungicides increased from 2.5 million units in 1990 to 4.3 million units in 1994. This increase can be explained by an increase in the banana growing area in Costa Rica and the need to increase fungicide applications per hectare because of fungal pathogens becoming less susceptible to fungicides.

All data above include technical and formulated material. Variation in the imported volumes of pesticides as documented in official import statistics may be different from the variation at the active ingredient level, i.e. importing a given quantity of an active ingredient as concentrated technical material will lead to lower numbers in import statistics than importing the same quantity of an active ingredient as a formulated product. Therefore, import data can only be interpreted as an estimate for actual pesticide imports.

In 1993, about 18% of all pesticide imports (in volume terms) belonged to the categories Ia (extremely hazardous) and Ib (highly hazardous) and 24% in the WHO category II (moderately hazardous). WHO categories do not apply to

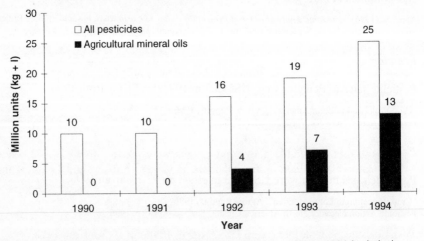

Figure 3. Quantities of pesticides imported to Costa Rica from 1990 to 1994 (technical material and formulated products in million units (kilograms and litres)) (source: Cámara de Insumos Agropecuarios, adapted from Agne (1996)).

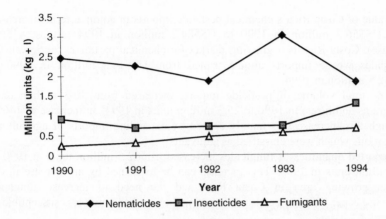

Figure 4. Volumes of fumigants, nematicides and insecticides imported to Costa Rica from 1990–1994 (in million units (kilograms and litres)) (source: Cámara de Insumos Agropecuarios, adapted from Agne (1996)).

fumigants and therefore many harmful pesticides applied as fumigants are included in the unclassified category (10%). WHO categories presuppose judicious and safe use of pesticides.

In 1994, the USA was the biggest supplier of pesticides, covering 37% of Costa Rica's pesticide imports, followed by Switzerland (17%), Germany (14%) and Colombia (11%)

Agne, S. (1996) *Economic Analysis of Crop Protection Policy in Costa Rica.* Pesticide Policy Project Publication Series No. 4, September 1996, pp.31–32. University of Hannover

Ecuador

Pesticide sales in Ecuador were US$93.3 in 1994, up 13.7% from 1993.

"Crop Protection in Latin America", Agrow Reports, 1996
Pesticide Action Network North America Updates Service, April 16, 1996

Peru

Pesticide sales in Peru were US$84.3 in 1994, up 27.2% from 1993.

"Crop Protection in Latin America", Agrow Reports, 1996
Pesticide Action Network North America Updates Service, April 16, 1996

Markets: Asia

China

Estimates of pesticide use have varied widely but it appears that active ingredient totals doubled between 1986 and 1991 from 160,000 tonnes to more than 300,000 tonnes.

"Agriculture in China." T.C. Tso, paper presented at The International Conference on Integrated Resource Management for Sustainable Agriculture, Beijing, September 1993.

Chinese market accounts for 5% of global pesticide sales, value of possibly US$1000 million. Since 1990, China has been the world's second largest agrochemical producer, a predominantly national industry remaining protected by the State. Some agrochemical dealers believe sales figures are actually higher than quoted above (as much as US$2 billion). They predict annual increases of 5% between 1996 and 2005. It is also thought China will increase its use of higher value, lower toxicity products.

Herbicide use is increasing, according to some sources, by 2 million ha per year as a result of changes in farming and cropping practices. Insect resistance resulting from heavy use of pesticides in cotton has led to spray applications escalating from 8–10 to 15–25 applications per season over 5 years to 1997. More than other factors, insect resistance is encouraging Chinese farmers to adopt IPM although with heavy reliance on *Bacillus thuringiensis* (*Bt*).

Grimes, A., *Crop Production Opportunities in China,* Report DS147, Agrow, PJB Publications
Pesticides News 39, March 1998, page 14

India

Insecticides dominate the Indian agrochemical market, with a share of about 74%. Farmers are using more pyrethroids but the demand for organophosphates is decreasing. Herbicides and fungicides account for about 12% each and in recent years, both have been increasing their market share by about 1% annually.

Agrow, *World Crop Protection News*, 1 January 1999

India is one of the few remaining countries still engaged in the large scale manufacture, use and export of some of the most toxic chlorinate pesticides, such as DDT, HCB and pentachlorophenol.

Santillo, D., Johnston, P., Stringer, R. and Edwards, R., A catalogue of gross contamination: Organochlorine production and exposure in India, *Pesticides News* 36, June 1997, page 4

India is one of only two countries world-wide (along with the USA) to have applied more than 100,000 tonnes of DDT since its initial formulation.

Voldner, E.C. and Li, Y.-F., Global usage of selected persistent organochlorines. *The Science of the Total Environment*, 1995, 160/161:201-210

Approximately 125 companies manufacture more than 60 technical grade pesticides in India, including "Dirty Dozen" pesticides BHC, DDT and methyl parathion, with an estimated total production capacity of 126,000 tonnes. In the 12 months leading up to March 1996, companies in India produced 86,000 tonnes of pesticides and the volume of pesticides sold in India rose by 5% to 83,400 tonnes, including 18,000 tonnes of HCH. Insecticides were 67% of the market, fungicides 22% and herbicides 10%. The next year to March 1997 saw a growth of 7.5% in value of the Indian pesticide market to US$602 million.

Agrow, *World Crop Protection News*, July 12 and September 27, 1996 and September 12 and October 3, 1997.

Table 3. Estimated Indian production of selected pesticides (tonnes).

Pesticide	1994/95	1996/97
BHC	32,000	20,000
DDT	4,300	4,400
Endosulfan	6,700	7,000
Malathion	2,800	4,000
Mancozeb	4,100	4,200
Monocrotophos	8,000	10,000
Methyl parathion	2,100	2,400
Phorate	4,100	4,100

Source: Agrow, *World Crop Protection News*, July 12 and September 27, 1996 and September 12 and October 3, 1997.

Pakistan

In 1997, 44,872 tonnes of pesticides were imported into Pakistan, an increase of 3.8% on 1996, according to figures from the Department of Plant Protection. However, the value of the market remained static at US$225 million.

Insecticides accounted for 74% of the total agrochemical market value. Within this sector, organophosphates have 41% of the market by value, pyrethroids 20%, organochlorines 6% and carbamates 5%. The remainder of the agrochemical market was represented by herbicides 14%, fungicides 9%, acaricides 2% and fumigants 1%.

Seventy-six per cent of the pesticides market is for use on cotton and the cotton area was 9.1 million hectares in 1997/98. 4% is used on rice, 3% on sugar cane and 17% on wheat and other crops.

Pakistan's manufacturing capability has declined since the 1970s as the country has become more import oriented. Liberalisation of the import policy in 1994 led to a 74% increase in imports to 43,374 tonnes in 1995.

"Pakistan agrochemical usage rises" Agrow, *World Crop Protection News*, December 11, 1998

The sale of pesticides in 1995 was US$222 million (not including the large quantities of pesticides smuggled across the border). Pyrethroids have 45% of the market by value, organophosphates 39%, organochlorines 9% and carbamates 4%.

The Agriculture Census shows that the proportion of farms using chemical pesticides has increased from 4% in 1980 to about 25% in the 1990s, that is 1.28 million farms or up to 16% of the total cropped area.

According to the Prime Minister's Task Force on agriculture, approximately 90% of the insecticides are used on cotton crops, which means that 2.68 million hectares are the target of pesticide use.

Recent legislation reflects a concern with the adulteration of pesticides rather than with the quantities of pesticides used.

"Use of pesticides in Pakistan", *Pesticides News* 37, September 1997, page 5

Consumption of pesticides increased from 3677 million tonnes in 1981 to 20,279 million tonnes in 1993.

"Round-up of pesticide regulation in selected countries of Asia", *Agrochemical News in Brief,* Vol. XIX, No. 3, July–September 1996

South Korea

Between 1980 and 1995, pesticide use increased from 16,132 tonnes to 25,834 tonnes, an increase of approximately 63%. Much of this growth was due to insecticide use on fruits, vegetables, ornamentals and greenhouse crops.

Pesticide Action Network North America Updates Service, November 8, 1996

Thailand

Thailand is a major market for pesticides with an annual growth rate between 1982 and 1992 of 8.8%. Since 1992, the market has continued to grow but at a slower rate. In 1994, sales totalled US$247 million. Following rapid growth in recent years, the herbicide market is now 51% of sales, while insecticides are 38% of the market and fungicides are 10%.

As most of the pesticides used in Thailand are imported, the large increase in pesticide imports during the period 1976 to 1995 (see Fig. 5) is closely related to increases in pesticide use. Most pesticides are imported with foreign companies possessing the biggest market share. 63% of the pesticides imported in 1992 fell into the WHO categories Ia (extremely hazardous) and Ib (highly hazardous).

Jungbluth, F. (1996) *Crop Protection Policy in Thailand. Economic and Political Factors Influencing Pesticide Use*. Pesticide Policy Project Publication Series No. 5, December 1996, pp.29-33. University of Hannover

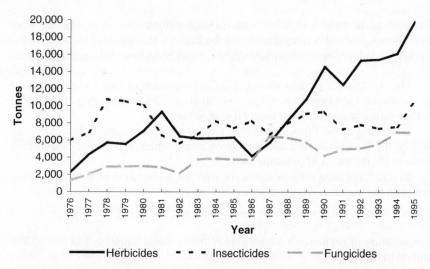

Figure 5. Quantity of pesticide imports in Thailand (1976 - 1995) (source: Regulatory Division: Pesticide statistics. Various issues, Agricultural Regulatory Division, Department of Agriculture, Bangkok, Thailand. (Adapted from Jungbluth, 1996)).

Vietnam

Agrochemical sales rose by 21.8% from $89.5 million in 1996 to $109 million in 1997. This large increase was due to an increase in cultivated area and high pest pressure, leading to a greater demand for fungicides and herbicides. The majority of the Vietnamese market is insecticides at 47.5% of the 1997 sales, while sales of herbicides and fungicides accounted for 25.2% and 24.3% respectively. The government-owned Vietnam Pesticide Company, which had 30% of the local market in 1996, says that pesticide use is still lower than demand.

Agrow, *World Crop Protection News*, September 18, 1998

Markets: Africa

Kenya

The Kenyan pesticide market was approximately US$40.4 million in 1992, placing Kenya among the highest pesticide users in sub-Saharan Africa according to a report issued by the WWF. Approximately 60% of pesticides used annually are applied to coffee.

All pesticides used in Kenya are imported, averaging out at 7300 tonnes annually according to the Kenyan Pest Control Products Board, although there are large fluctuations between years, the smuggling of pesticides is a problem and the records fail to distinguish between formulated product and technical grade concentrate which is imported for formulation in the country. The majority of pesticides are imported from Europe and the US. Twenty-two per cent of the pesticide volumes imported are classified by the WHO as extremely and highly hazardous (Class Ia and Ib) while moderately hazardous (WHO Class II) pesticides make up 20% of the imports. Many of the pesticides exported are not registered for use in the country of origin.

"The cost of hazards posed by pesticides in Kenyan export crops" *Pesticides News* 29, September 1995, page 6

Madagascar

Between 1947 to 1982, approximately 35,000 tonnes of chemicals were used on rice, cotton, tobacco and sugar cane.

Between 1986 to 1994, an average of 540 tonnes of pesticides per year were imported. Twenty per cent reached local markets through donor contributions (mainly Japan).

In 1992, the locust control campaign attracted donation of 19,000 litres of fenitrothion (GTZ), 5000 litres of diazinon and 40,000 litres of lambda-cyhalothrin (USAID)

Use on cash crops is high; pesticides account for nearly 20% of cotton production costs.

In 1994, it was estimated that more than 60% of pesticides sold in the country are partly or completely altered.

Pesticides News, IPM in Madagascar Supplement

Senegal

In 1992, US$2200 million were spent on pesticides, the majority for use on cotton.

Senegal Case Study, Pesticides Trust 1993

Pesticide Exposure and Health Effects

Many pesticides that have been banned or whose use has been severely restricted in industrialised countries are still marketed and used in developing countries. These chemicals pose serious risks to the health of millions of farmers and the environment.

FAO Director-General, Dr Jacques Diouf

Many developing countries do not have effective monitoring systems in place to assess the extent of pesticide poisonings and the majority of cases are unreported; the WHO estimated that there are 50 cases of poisoning for every case reported. The most recent estimate in 1990 by the WHO of the number of cases of acute occupational pesticide poisonings is 25 million per year world-wide. With the continued increase in use of pesticides, it is to be expected that this figure will also increase.

50% of all pesticide related illness and 72.5% of recorded fatal pesticide poisonings occur in developing countries, although these countries account for only 25% of the pesticides used world-wide.

Knirsch, J. "Pestizide und Dritte Welt — Mehr als nur win Kreislauf der Gifte", in Pestizid Aktions Netzwerk (PAN) (ed.), *Pestizid in Lebensmitteln*, PAN Germany

Large quantities of chemical pesticides considered extremely or highly hazardous by the WHO are being imported or manufactured and used in developing countries. Many of these pesticides are then applied by people with little or no training in safe application or storage. Studies of farmers and their families repeatedly show that there is a high risk of exposure to toxic pesticides through lack of protective clothing, leaking spray equipment, mixing and application of pesticides with bare hands, and storage of pesticides with food.

Some anecdotal accounts are included in the following examples of pesticide poisonings in order to supplement the comparatively scarce number of studies and official records and to cover as many countries as possible; for some countries, anecdotal accounts were the only source of information that could be found.

Latin America

Latin American farmworkers are 13 times more likely to suffer pesticide poisoning than US farmworkers.

Tansey, R.T. *et al.* (1995) "Eradicating the Pesticide Problem in Latin America", *Business and Society Review*, 92:55–59

Brazil

It is estimated that 18% of all new Wilms' tumours, affecting the kidneys of Brazilian children, are attributable to pesticide exposure of their parents. Consistently elevated risks were seen among children whose father or mother carried out farm work which involved the frequent use of pesticides. As elsewhere, many pesticides are used in Brazil and researchers estimated that 73 insecticides, 42 fungicides and 59 herbicides are in use. Atrazine and dichlorvos are particularly widely used and both are rated as possibly carcinogenic by the International Agency for Research in Cancer.

Sharpe, C., Franco, E. *et al.*, Parental exposure to pesticides and the risks of Wilms' tumour in Brazil, *American Journal of Epidemiology*, 1995, 141:210-217
Pesticides News 28, June 1995, page 25

According to Brazil's Ministry of Health, 6000 cases of pesticide poisoning were reported in 1993. Based on the WHO's estimate that there are 50 cases of poisoning for every case reported, the Servico Brasileiro de Justica e Paz (SEJUP) estimated that as many as 300,000 people are poisoned in Brazil every year and that this number has been rising annually.

Health experts reported to the Rio Grande do Sul State Legislature that poisoning by widely used pesticides in agricultural regions can lead to physical and mental problems including anxiety, irritability, loss of memory and depression and these symptoms could lead to suicide. In Arapiraca, a municipality in the state of Algoas, the rate of suicides was 16 per 100,000 in 1996 — five times the rate for the state. Most committed suicide by drinking agrochemicals and all either used agrochemicals in their jobs or lived in an area where chemicals were used.

SEJUP News, November 20 and December 5, 1996
"Poisonings in Brazil" *Global Pesticide Campaigner*, Vol.7 No.1, March 1997

Chile

The Chilean fruit export business is very important to the national economy and these producers have responded to the requirements of the Western market for "quality" and quantity. In turn this has led to an increase in the pesticides used. Many of the pesticides used are WHO class Ia "extremely hazardous". The main exposure route for farmworkers is dermal. Pregnant women who continue to work in fields sprayed with pesticides run an increased risk of exposing their unborn children. The apple and pear industries are highly manual with the fruits being picked, sorted and packed by hand.

Newbold, J., "Chile pays the price for exports" *Pesticides News* 37, September 1997, Page 8

OP insecticides and herbicides dominate the market in Latin America and according to a WHO report, 10–30% of farmworkers tested showed significant cholinesterase inhibition, an important biomarker of exposure to OPs.

World Health Organization and United Nations Environment Programme, Public Health Impact of Pesticides Used in Agriculture, WHO Geneva, 1990.
Pesticides News 37, September 1997, Page 8

For some years the health services in the VI, VII and VIII regions have detected an increase in the number of birth defects, cancer and other diseases among agricultural workers and their children. In 1993, 44 children were born with congenital malformations in the Curicó Regional Hospital. The same thing happened to a large proportion of children in 1994. Almost all the parents of these children had been exposed to pesticides due to the fact that they worked in fruit orchards, packing plants or lived near them.

Rozas, M.E., "Pesticides in Chile", Institute of Political Ecology, Chile 1995
Pesticides News 37, September 1997, Page 8

Colombia

At least 60 people were poisoned and one died in 1993 in an incident when endosulfan was used on coffee crops.

Pesticide Action Network North America Updates Service, July 28, 1993

1. The Flower Industry
Up to 20% of commonly used pesticides in Colombia are banned or not registered in Europe and the United States. Colombia is the worlds second largest flower exporter after the Netherlands but to ensure that the flowers are not rejected by importing countries, Colombian flower farmers douse the plants in pesticides to prevent any disease or blemish. The result is poisoned workers, contaminated water and parched soil. Little empirical evidence has been collected examining the direct effects of pesticides on the health of workers in the Colombian flower industry. However, adding together workers' testimonies, research carried out on

pesticides in other agricultural sectors and individual cases shows how Colombian flower workers are suffering from working with dangerous pesticides.

On 28 and 29 December 1994, 13 adult patients entered San Pedro Claver Clinic in Bogata with a loss of strength, muscular weakness and tingling in their legs. The doctors diagnosed peripheral polyneuropathy. All the patients came from a flower farm where in the previous eight days they have been exposed to a product, Karate, whose active ingredient is lambda-cyhalothrin.

Carillo, S., Report to the Health Ministry, San Pedro Claver Clinic, Bogota

A recent study by the Finnish Institute of Occupational Health found that woman agricultural workers exposed to certain pesticides during the first three months of pregnancy had double the risk of giving birth to deformed children. Some of the chemicals used in the study—dichlorvos, aldicarb, mancozeb, captan and naled—are used in the flower industry and other agricultural activities in Columbia. More than 60% of flower workers in Colombia are women.

CACTUS, *Information Newsletter about the Flower Industry*, No. 1 Sept 1995

2. Workers' Testimonies

Women's experiences: "Once when they were fumigating, I went to leave greenhouse. The fumigator's hose pipe was broken, and I stepped on it and liquid squirted in my face. Although I washed myself immediately, I began to vomit and have a fever," said Elvira Rincon, a 51-year-old former flower worker. Ms Rincon spent nine days in hospital, with serious poisoning. Since then she has suffered a miscarriage, cancer of the womb and spinal problems. Ms Rincon blames her 20 years in the flower industry.

"The work is really hard because the Green houses are hot. It gives you a headache and you feel dizzy. They fumigate and you're right there working," said Florangela Campos, who used to work on the flower farms.

Fumigators: "Once I had a pair of waterproof trousers which were ripped all the way down the inner seam, so I had to get a stapler and staple them together," said Cesar Campos, a former flower worker. "Wearing a mask in the greenhouse would get too hot, so sometimes I took off the waterproof clothing and by the end my T-shirt would be soaked. As you walk along next to the flowers, they are soaked in pesticides and they brush against your skin. The woman would return sometimes just after I'd fumigated and they would immediately start touching the flowers again. I always felt a bit sick when I was fumigating."

Matheson, M., "The Colombian flower trade—success at a price" *Pesticides News* 32, June 1996, pages 3–5

Costa Rica

Statistics from the National Institute of Insurances in Costa Rica indicate that aldicarb was the number one cause of pesticide poisonings in the banana producing region of Guaypil in 1988, accounting for 113 poisonings and over 30% of all cases linked to a specific pesticide.

In 1986, during the first week that aldicarb was used on Standard Fruit Co.'s banana plantations in Rio Fro, Costa Rica, more than 100 workers were poisoned by the pesticide.

Seidenburg, K., "High levels of Aldicarb found in bananas", *Global Pesticide Campaigner*, June 1991, Vol. 1, No. 3.

Researchers in Costa Rica have analysed pesticide related injuries reported in the Guàpiles region to the National Insurance Company in 1993 and 1996 where about 13,000 people work in agriculture. The results showed that pesticide related injuries occurred 5.2 and 3.4 times per 100 agricultural workers in 1993 and 1996 respectively. Injuries most frequently involved herbicides usually resulting in eye or skin lesions (1.6 and 1.0 per 100 agricultural workers for 1993 and 1996). Incidence of health effects with insecticides and nematicides were also high with 0.7 and 1.0 cases occurring per 100 workers, mostly involving OPs and carbamates. This compares badly with US state of California where the annual reported pesticide poisonings is 0.249 per 100 workers. The reduction in 1996 might be due to changes in the pesticides used, although a decrease in the number of workers covered by the National Insurance Company may have led to under reporting.

VanWendel de Joode, B.N., and Wesseling, C., Pesticide related occupational injuries in Costa Rica - a comparison between 1993 and 1996. The International Commission on Occupational Health's 12th International Symposium on Epidemiology in Occupational Health, Zimbabwe, 16 - 19 September.

More than 1000 workers in banana plantations have become sterilised as a side-effect of applying DBCP (Thrupp, 1989). A more recent estimate revises this to as much as 10,000 workers sterilised (Schonfield *et al.* 1995).

Subsequent severe psychological repercussions of being sterilised include impotency, severe depression, despondency, grief and confusion. The men are subject to social isolation and ridicule. For many it has resulted in separation, divorce and loss of jobs (Thrupp, 1989)

Occupational pesticide poisonings have been considered a serious problem in Costa Rica for many years (e.g. Thrupp, 1989). The number of registered poisoning cases can be assumed to be lower than the actual number of poisonings as monitoring is based on voluntary information provided by poison victims and physicians.

There has been an almost constant increase in registered poisonings between 1980 and 1994 from 593 to 1144. In 1994, >99% of all agrochemical poisonings were caused by pesticides (Fig. 6).

Children under 5 years are highly affected, possibly due to the lack of adequate storage facilities for pesticides on the farmers' property (Fig. 7). The major risk group is men between 15 and 44, who represent the largest part of the agricultural working force.

In 1993, 42 lethal intoxications occurred, no data are available for 1994.

Thrupp, L.A., "Direct Impact: DBCP poisoning in Costa Rica". *Dirty Dozen Campaigner* May 1989
Schonfield, A., Anderson, W. and Moore, M., "PAN's Dirty Dozen Campaign — the view at ten years". *Global Pesticide Campaigner*, Vol. 5 No. 3, September 1995
General source: Agne, S. (1996) *Economic Analysis of Crop Protection Policy in Costa Rica.* Pesticide Policy Project Publication Series No. 4, September 1996, pp. 31-32. University of Hannover

Table 4. Analysis of registered pesticide poisoning cases in Costa Rica in 1994.

Means of intoxication with agrochemicals	
Ingested	48%
Inhaled	29%
Absorbed through the skin	11.4%
Inhaled and absorbed through the skin	10%
Classifications of registered poisonings	
Occupational	34%
Accidental	43%
Suicidal	19%
Gender split of poisonings	
Male	70%
Female	30%

Source: Agne (1996).

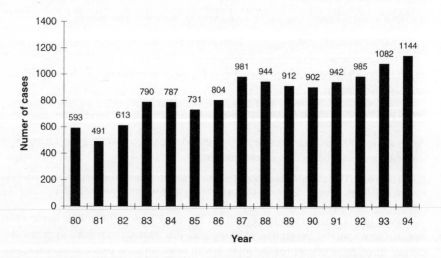

Figure 6. Agrochemical poisoning cases in Costa Rica registered at the National Centre for Poisoning Control from 1980 to 1994 (source: Centro Nacional de Control de Intoxicaciones, San José, Costa Rica).

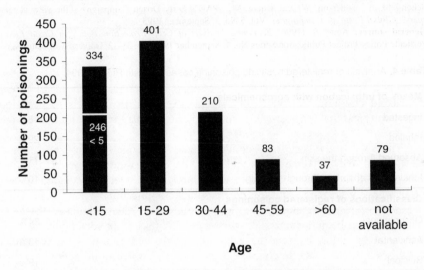

Figure 7. Distribution of pesticide poisonings in 1994 according to age (source: Centro Nacional de Control de Intoxicaciones, San José, Costa Rica).

Guatemala

The estimated number of pesticide poisoning cases is estimated to be in the range of 11,000–30,000 per year.

"Problems persist in Central America" *Pesticides News* 41, September 1998, page 11

Mexico

The Mexican Ministry of Health reported a total of 1500 deaths from pesticide poisonings in 1993.

Diaz-Ramo, P. (1994) "Huicholes and Pesticides", video, Patricia Diaz-Ramo Productions, Mexico

Among the 200 pesticides authorised by the Mexican government for agricultural use, 32 have been banned in other countries.

United Nations, (1994) Consolidated list of products whose consumption and/or sale have been banned, withdrawn, severely restricted or not approved by governments. New York, USA

A survey of 500 farmers in the Fraylesca region, Chiapas found that farmers did not use any specific protective clothing and that dermal exposure to pesticides during spraying accounts for the main health risks. In 22% of the observed cases, knapsack sprayers leaked and there was visible drenching of the clothes (20%), the skin (24%) and the hands (48%) with the pesticide solution. Work clothes are often worn for several days in a row before being washed. A four year intervention programme found that communication and training had a favourable impact on comparatively simple, cheap safety practices used by farmers, while the more cumbersome practices were not employed for long.

Atkin, J. and Leisinger, K.M. (2000) *Safe and effective use of crop protection products in developing countries.* CAB International, UK

A study in the Yaqui Valley of Sonora, Mexico, comparing two groups of 4- and 5-year-old children who were very similar except in their levels of pesticide exposure found that long-term exposure to pesticides may impair children's brain function.

The two study groups of children share the same genetic and cultural background, eat the same foods and drink the same water. However, the first group live in the valley, a farming area with relatively intense pesticide use; two crops per year and up to 45 pesticide applications per crop using organophosphates, organochlorines and pyrethroids. Contamination of the local population has been documented, with women's breast milk containing levels of lindane, heptachlor, benzene hexachloride, aldrin and endrin all above limits established by the FAO. In contrast, the second group live in the foothills where most families are involved in ranching and pesticide use is minimal.

The children living in the valley were found to have significantly less stamina and hand–eye co-ordination, poorer short-term memory and were less adept at drawing a person than those children from the foothills.

An anthropological approach to the evaluation of preschool children exposed to pesticides in Mexico. *Environmental Health Perspectives*, 106:6, June 1998. Summarised from: Pesticide Action Network North America Updates Service, June 5, 1998

Nicaragua

In 1996, 1363 cases of pesticide poisoning were reported through the official health system. The actual number of cases is estimated to be about 10,000 per year.

"Problems persist in Central America" *Pesticides News* 41, September 1998, page 11

In 1987 there was an epidemic of 584 pesticide poisonings of which metamidophos and carbofuran were responsible for 77%.

Pesticide Action Network North America Updates Service, May 19, 1995

A survey of 36 men who suffered an accidental, work related organophosphate poisoning were found to perform significantly less well at a series of neuropsychological tests (including a battery of WHO tests) than a control group.

Rosenstock, L. *et al.* (1991) Chronic central nervous system effects of acute organophosphate pesticide intoxication. *The Lancet* 338, 223-227

Paraguay

A 1990/91 study of paralysis of the limbs in children, originally thought to have been caused by polio, suggested that drift from nearby cotton fields of the organophosphate pesticide, monocrotophos was the most likely culprit.

Pesticide Action Network North America Updates Service, May 19, 1995

Middle East

Between 1967 and 1970, there were 874 recorded poisonings with 26 fatalities linked to endrin-contaminated flour. There have been at least 400 deaths from a single case of dieldrin-contaminated water.

Schonfield, A., Anderson, W. and Moore, M., "PAN's Dirty Dozen Campaign — the view at ten years." *Global Pesticide Campaigner*, Vol. 5 No. 3, September 1995

Asia

China

Pesticide poisoning is a major problem, resulting in large numbers of deaths; there were 10,000 farm worker deaths from pesticides in 1993.

In 1995, of the 15,300 pesticide poisoning cases reported in 27 provinces as a result of agricultural use, the majority were caused by organophosphate insecticides.

Grimes, A., *Crop Production Opportunities in China*, Report DS147, Agrow, PJB Publications *Pesticides News* 39, March 1998, page 14

1. OP Poisoning

In 1995, the National Statistics Bureau in China reported 49,377 pesticide poisoning cases (accidents and suicides) in 27 provinces, including 3204 deaths. Of these, 15,000 cases occurred during occupational pest control activities of which 91% involved OPs and a fatality rate of 0.5% (75 cases). The reporting system was only set up in 1992 and needs improving; some areas have not established a reporting system while in other areas reporting is limited. It is estimated that there could be a level of under-reporting by as much as 60% or more.

Chen, S. and Yao, P., Heavy OP poisoning toll in China, *Pesticides News* 32, June 1996

In 1995, OP contaminated vegetables were responsible for 200 students in Guangxi requiring hospital treatment.

Grimes, A., *Crop Production Opportunities in China*, Report DS147, Agrow, PJB Publications *Pesticides News* 39, March 1998, page 14

In 1987, the Health Department found that an outbreak of 120 cases of acute organophosphate poisoning in Hong Kong was caused by metamidophos residues on vegetables imported from China. Since 1987, another 600 poisonings have been attributed to the same cause.

Pesticide Action Network North America Updates Service, May 19, 1995

In 1987, the Chinese Ministry of Agriculture estimated that 10,000 people die per year from pesticide poisoning.

"Pesticides kill up to 10,000 people per year." Reuters News Service, June 24, 1987

In the first nine months of 1991, there were 101 cases of pesticide residues on vegetables poisoning 2086 people reported in the Guangdong province.

Thiers, P., Pesticides in China. *Global Pesticide Campaigner*, Vol 4 No 1, March 1994

India

A study on pesticide poisoning found that out of 635 pesticide poisoning cases, only 189 were reported to the hospital. About 25% of these cases were accidental and pesticides were the main toxic agents. The study uncovered farmer practices that included exposure to spray drift and use of the mouth to siphon pesticides from the container.

Saloke, V.M. (ed.), *Safe and Efficient Application of Agro-chemicals and Bio Products in South and South-East Asia,* proceedings of the International Workshop,28-30 May 1997, Asian Institute of Technology, Thailand
Pesticides News 38, December 1997, Page 16

In April 1990, over 100 people died after attending a wedding feast in Northern India when lindane powder was mistakenly added to the flour of a wedding meal.

Schonfield, A., Anderson, W., and Moore, M., "PAN's Dirty Dozen Campaign — the view at ten years." *Global Pesticide Campaigner*, Vol 5 No 3, September 1995

A survey of farmers' spray practices in Coimbatore found that use of protective clothing such as gloves when mixing pesticides and a face mask, full shirt and trousers when spraying was low and despite some improvement during a communication and training programme, their use fell the year after the end of the programme and it was concluded that the practices were not likely to be sustained. The survey also concluded that in areas of economic deprivation, a farmer will not spend money on protective clothing and will only value their health when they feel financially secure. If it is provided free or as a subsidised rate, a farmer might use it but if it hampers productivity at all, he will discard it.

Atkin, J. and Leisinger, K.M. (2000) *Safe and Effective Use of Crop Protection Products in Developing Countries.* CAB International, Wallingford, UK

Indonesia

In 1995, a survey of 214 farmers who spray pesticides on vegetable crops found 69 different signs and symptoms affecting the health of sprayers during the spray season. Symptoms were only recorded if they arose during or within a few hours of the spray operation. The most frequently observed are shown in Table 5.

Table 5. Frequency of symptoms experienced by a group of 214 farmers in Indonesia after spraying pesticides.

Symptom	Attack rate (%)
Fatigue	60
Muscle stiffness	54
Dry throat	30
Muscle weakness	23
Dizziness	21
Difficulty in breathing	18.5
Insomnia	17
Blurred vision	15.5
Stinging eyes	15
Flushed face	14
Chest pain	13.6
Headache	13
Salivation	13
Nausea	11
Itchy skin	9

Jishi, M. and Hirschorn, N., Relationship of pesticide spraying to signs and symptoms in Indonesian farmers, *Scandinavian Journal of Work and Environmental Health*, 1995, 21:124-33
Pesticides News 28, June 1995, page 25

Official records in Indonesia do not indicate a pesticide poisoning problem, but local studies estimate 30,000 cases per year with 2400 requiring hospitalisation.

Acute pesticide poisoning: a major global health problem. *World Health Statistics Quarterly* No. 43, 1990
Global Pesticide Campaigner, June 1991, Vol. 1, No. 3

Malaysia

Between 1988 and 1993, the number of persons admitted to hospitals due to pesticide poisonings were about 1300 per year, of which about 400 died. In the period 1988–1991, paraquat accounted for 60–70% of these admissions, and for as much as 90% of the deaths due to pesticide poisoning.

"Developmental Prejudice", *Pesticides News* 41, September 1998, page 10

Paraquat poisoning is extremely common in Malaysia, where it accounted for 66% of the 1442 reported pesticide poisoning cases between 1978 and 1985. A 1985 Department of Agriculture survey found that only 11% of the workers interviewed had been trained before handling paraquat while 67% did not receive any protective clothing from the plantation management. Thus it is not surprising that 64% of the workers reported symptoms of poisoning.

Sahabat Alam Malaysia "Workers campaign against paraquat". *Dirty Dozen Campaigner*, September 1989

In a 1991 study, every surveyed pesticide spray applicator working regularly with dimethoate reported suffering "often" from nausea, sore eyes and headaches, the symptoms of organophosphate poisoning. Other studies have indicated that this insecticide can cause anxiety and depression in people who have been regularly exposed.

Pesticide Action Network North America Updates Service, May 19, 1995

A study of 103 tobacco workers on 50 family farm units found that the OP, metamidophos, WHO Toxicity Class Ib (highly hazardous) was used on 96% of farms and was always applied using a knapsack sprayer; 46% of the knapsack sprayers observed in use were leaking. A third of the workers had two or more symptoms consistent with pesticide poisoning.

Cornwall, J.E. *et al.* (1995) Risk assessment and health effects of pesticides used in tobacco farming in Malaysia. *Health Policy and Planning* 10, 431-437

Health and safety practices of farmers need improvement: only 40% use gloves when mixing pesticides, 24% did not change clothes if wet with pesticides, 13% blow out clogged nozzles.

Saloke, V.M. (ed.), *Safe and Efficient Application of Agro-chemicals and Bio Products in South and South-East Asia,* Proceedings of the International Workshop, 28-30 May 1997, Asian Institute of Technology, Thailand
Pesticides News 38, December 1997, page 16

Pakistan

A UN report estimated that approximately 500,000 people a year are poisoned by pesticides in Pakistan of which 10,000 die.

Dawn, 26 August 1998

A survey of three hospitals found 52 people treated for non-suicidal pesticide poisonings.

Saloke, V.M. (ed.), *Safe and Efficient Application of Agro-chemicals and Bio Products in South and South-East Asia,* Proceedings of the International Workshop, 28-30 May 1997, Asian Institute of Technology, Thailand
Pesticides News 38, December 1997, page 16

Philippines

A survey in the Philippines found that 22% of the insecticides sprayed on rice are WHO Class Ia (extremely hazardous), primarily methyl parathion, while another 17% were WHO Class Ib (highly hazardous).

Heong, K.L. (1994) An analysis of insecticide use in rice: case studies in the Philippines and Vietnam. *International Journal of Pest Management* 40(2)

Government hospitals reported 4031 cases of pesticide poisoning between 1980 and 1987 (*Philippine Daily Enquirer,* June 1, 1994). Another report by the National Poisons Control and Information service listed 1302 poisoning cases between January 1992 and March 1993 in the National Capital region alone.

Hickey, E., "International citizens' campaign targets Hoechst Pesticides". *Global Pesticide Campaigner,* Vol.4 No. 3, September 1994

Most farmers have not received training in using pesticides or application equipment. While they generally perceive pesticides as hazardous and know that contamination should be avoided, protective measures are rarely taken. In one survey, 40% of farmers wore a hat, 35% wore a "mask" (normally a handkerchief) and 50% long sleeve shirts and trousers. Gloves and boots were not generally worn. Knapsack sprayers are widely used and the survey concluded that applicators face a high level of pesticide exposure, particularly dermal.

Saloke, V.M. (ed.), *Safe and Efficient Application of Agro-chemicals and Bio Products in South and South-East Asia,* proceedings of the International Workshop,28-30 May 1997, Asian Institute of Technology, Thailand
Pesticides News 38, December 1997, Page 16

Insecticides accounted for 21 out of 70 cases of aplastic anaemia admitted to the Philippine General Hospital, and 12 of the 21 were farmers.

Giongco-Baylon, H.V., Domingues, C.E., Perez, V., Lu, J. and Ona, N. (1982) Study of aplastic anaemia at the Philippine General Hospital. Manila. Unpublished paper

Comparisons of farmers with long-term exposure to pesticides in rice production with farmers with no history of exposure showed that the magnitude of chronic health effects and health costs are directly related to pesticide exposure and that the net benefits of insecticide use are negative.

Significantly increased occurrences of eye irritation (pterygium, 67% / 10%), skin effects (nail pitting, eczema, 45% / 0%), respiratory tract effects (45% / 23%), cardiovascular effects (49% / 46%), chronic gastritis (9% / 0%), kidney (26% / 15%) and haematological problems (89% / 74%) were observed amongst the exposed farmer group (57 farmers) over the control group (38 farmers).

Pingali, P.L. *et al.* (1994) Impact of pesticides on farmer health: a medical and economic analysis. *Rice Pest Science and Management,* International Rice Research Institute, pp.277-289.

Analyses of mortality records in rural areas as pesticides were introduced during the "green revolution" (1960s and 70s) showed that there was a significant increase in mortality due to toxicosis and "unexplained death" during this time amongst men (who did most of the spraying) in comparison with women.

Loevinsohn, M.E. (1987) Insecticide use and increased mortality in rural Central Luzon, Philippines. *The Lancet* (June 1987), 1359-1362

Thailand

In Thailand, the Division of Epidemiology of the Ministry of Public Health has the primary responsibility of collecting poisoning data. However, these data rely on case reports of Government Hospitals and some private clinics and therefore the number of actual poisoning cases are assumed to be understated (Sinhaseni, 1990). A survey about poisoning cases among agricultural workers concluded that only 2.4% of workers in poisoning incidents consult a hospital (Wongpanich, 1985). Jungbluth (1996) estimates that the total number of poisoning cases in Thailand is 39,600 with total health costs adding up to about 13 million baht ($520,000) (see box below).

Official data from the Epidemiology Division shows the number of occupational poisoning cases occurring in 1994 was 3165. In a study conducted by Whangthongtham (1990), health costs have been assessed for poisoning cases in Pathum Thani, Thailand. According to this survey, 25% of poisoning cases are treated in hospitals, 52% in private clinics and 23% in health offices. The costs related to these treatments are 550 baht for hospitals (3 days treatment), 120 baht for clinics and 70 baht for health offices. Additionally, labour costs in the form of lost labour days have to be calculated. The costs per labour day are calculated with 100 baht per day, the loss of labour days amounts to 3 days for hospital treatment, and 0.5 days for both clinic and health office treatment. Relating the poisoning cases to the average costs from medical treatment and lost labour days assessed in that survey to be 328.5 baht, the implied total health costs therefore amount to about one million baht. If we consider that the available statistics underestimate the actual poisoning incidents and that the death cases are not included, the calculated costs may serve as the lower boundary of the actually implied health costs.

To conclude to a more realistic amount of the health cost assessment results of the study of Whangthongtham (1990) are used to calculate poisoning cases in relation to insecticide market volume. The poisoning cases per hectare and the intensity of insecticide use are needed for this calculation. Firstly the reported poisoning cases are mainly due to insecticide use and are therefore related to the quantity of insecticides used. Secondly, poisoning cases are not location specific and finally, the hazardousness of the pesticides used is comparable for all crops.

If we consider, as indicated in the study, that 86% of the total poisoning cases of tangerine growers (total of 2121 cases) in Pathum Thani are caused by insecticides, the number of cases would amount to 1824. These cases are related to the tangerine growing area of Pathum thani (24,926 ha) and the intensity of insecticide use in citrus (US$235/ha). The derived poisoning cases per US$ insecticide use are then related to the total insecticide market in Thailand (US$93.5 million). The result would be 29,118 poisoning cases due to insecticides in Thailand per year.

Furthermore, considering the data from the Epidemiological Division, Ministry of Health in 1995 which shows that only 64% of all poisoning cases in Thailand are related to insecticides and 36% are related to herbicides and other pesticides, the calculated number of insecticide poisonings could be used to calculate the total number of pesticide poisonings. Consequently, 36% (10,482 cases) of poisonings due to other pesticides have to be added to the insecticide cases. Therefore, the total number of poisoning cases would amount to 39,600 cases. If these cases are weighted with the average health costs per poisoning case, total health costs sum up to about 13 million baht (US$520,000).

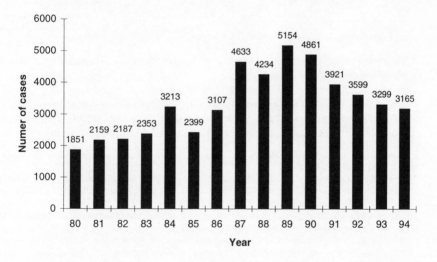

Figure 8. Occupational pesticide poisoning cases (1980 to 1994) (source: Epidemiological Division, Ministry of Health, 1995).

There is no apparent reason for the recent decrease in pesticide poisonings since the amount of pesticides imported and used has increased and no radical change in the type and hazardousness of pesticides used and the application technology chosen has taken place.

In the first seven months of 1996, Ministry of Public Health figures showed that 1760 people were hospitalised and 16 people died due to pesticide poisoning (Bagoglu, 1996).

Over 47% of all poisoning cases are based on organophosphate use, followed by herbicides (22%) and by the carbamate group (11%). No information on the long term effects of pesticide use is available.

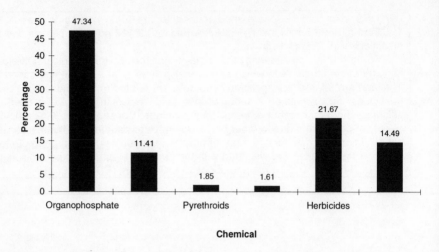

Figure 9. Pesticide poisoning by type of chemical (1994) (source: Epidemiological Division, Ministry of Health, 1995).

In a survey conducted by Khuankaew (1995), of the women questioned, 80% stated that they have been poisoned, reporting acute effects like dizziness, muscular pain, headache, nausea, weakness and difficulty in breathing.

Bagoglu, N. (1996) *Unternehmerbriefe - Thailand.* Bundesstelle für Aussenhandels informationen, October 1996, Köln, Germany

Khuankaew, O. (1995) *IPM and Women.* Report on Short Term Consultancy, Integrated Pest Management in Selected Fruit Trees Project, Thai-German Plant Protection Programme, Department of Agricultural Extension, Bangkok, Thailand

Sinhaseni, P. (1990) *Regional Pesticide Review - Thailand.* International Development Research Centre, Chulalongkorn University, Bangkok, Thailand.

Whangthongtham, S. (1990) Economic and environmental implications of two alternative citrus production systems - a case study from Pathum Thani Province. Masters Thesis, Asian Institute of Technology, Bangkok, Thailand

Wongpanich, M. *et al.* (1985) *Pesticide Poisoning among Agricultural Workers: a Research Report.* Mahidol University, Bangkok, Thailand

General source: Jungbluth, F. (1996) *Crop Protection Policy in Thailand. Economic and Political Factors Influencing Pesticide Use.* Pesticide Policy Project Publication Series No. 5, December 1996, pp 29-33. University of Hannover

Vietnam

A survey in Vietnam found that 17% of the insecticides sprayed on rice are WHO Class 1A (extremely hazardous), primarily methyl parathion, while another 20% were WHO Class 1B (highly hazardous).

Heong, K.L. (1994) An analysis of insecticide use in rice: case studies in the Philippines and Vietnam. *International Journal of Pest Management* 40(2)

Africa

Ethiopia

Ethiopian agriculture provides a livelihood for 80% of the population. State farms operate intensive production and account for the majority of the country's pesticide use. A study carried out on state farms in 1994–95 examined the health impact on 430 pesticide users (who use hazardous pesticides including regular use of OP's ranging from classifications of "moderately" to "extremely" hazardous by the WHO classification) against 161 unexposed workers from other occupations.

There were significantly lower levels of cholinesterase activity (low levels interfere with the function of the nervous system and are an indication of exposure to OPs) among workers on state farms confirming absorption of OP formulations. Awareness of pesticide hazards was extremely low, with only 12% of workers knowing the danger of poisoning and trying to take some precautionary measures. Some of the remaining 88% believed they were resistant to toxic formulations. Only 18% used protective gear while others commonly wore ordinary cotton work clothes.

Kibruyisfa Lakew and Yalemtsehay Mekonnen, A study among agricultural workers in Ethiopia, *African Newsletter on Occupational Health and Safety*, 1997, 7:68–70
Pesticides News 39, March 1998

Kenya

In 1993 and 1994, acetylcholinesterase levels in 666 Kenyan agricultural workers were investigated. Of these, 58.6% were mainly spray applicators and the remaining (276) 41.4% were unexposed controls. Acetylcholinesterase inhibition was found in all exposed individuals, with levels reduced by an average of 33%. The unexposed group had a non-significant decrease of only 4%. The dramatic acetylcholinesterase inhibition observed can lead to chronic clinical and sub-clinical intoxication.

Ohaya-Mitoko, G., *Occupational Pesticide Exposure among Kenyan Agricultural Workers*, Kenyan Medical Research Institute and Department of Epidemiology and Public Health, Wageningen Agricultural University
Pesticides News 38, December 1997 page 1

A survey interviewing 105 pesticide mixers, loaders and applicators on 19 coffee estates found that most workers reported experiencing poisoning, symptoms including skin irritation (84%), breathing difficulties (71%), stomach problems (58%) and nausea (20%). These symptoms occurred during periods of pesticide application and did not arise when processing coffee or weeding manually. Working practices for pesticide workers were poor, they mixed chemical concentrates using bare hands, 53% worked barefoot and 11% wore open slippers, 59% wore overalls, none of them had any training in using pesticides and equipment was generally poor with leaks occurring regularly. While the majority of workers were aware of the health consequences of pesticides, the fear of job loss led most to dismiss occupational safety as an unaffordable luxury.

Partow, H. (1995). "Pesticide Use and Management in Kenya" Institut Universitaire D'Etudes de Developpement, Geneva. *Pesticides News* 29, September 1995 pages 6–9

Madagascar

Farmers often use pesticides with little regard to safety and rarely have adequate information on the nature of pesticides or the risks involved. No study on the health, environmental and social impacts has been conducted although pesticide contamination, poisonings and lethal accidents have been reported at many sites.

Pesticides News, IPM in Madagascar, Supplement

Nigeria

Farming households in Nigeria may be exposed to pesticide hazards due to poor handling conditions and the use of highly hazardous pesticides, some of which are widely banned. In a survey of 180 households, 33% reported using lindane and 17% used DDT. Widely banned pesticides are all used because they are readily available or cheaper than pesticides recommended by the government extension agencies.

Pesticide knowledge was poor particularly regarding safe storage; pesticides were often stored alongside food and in living areas in non-pesticide containers such as food and beverage containers.

Safety periods between spraying and harvesting were not observed, with some households harvesting their produce only three days after the last application instead of the recommended five to 21 days depending on the pesticide used. Over 50% of the produce is for family consumption. Since family members work with pesticides in the fields and then consume much of the same produce at home, they are exposed to pesticides both occupationally and through residues on foods.

Udoh, A.J. "Safety research study", *Pesticides News* 40, June 1998, pages 8–9

Senegal

Pesticide containers are often reused by rural women to carry or store their crops. In 1983, 19 people died following a meal prepared with cooking oil that had been stored in a bottle that had contained parathion. The woman that prepared the meal committed suicide because she thought her couscous was responsible for the deaths. In another case, a mother accidentally killed her daughter by using an endosulfan based preparation to treat head lice.

Sow, M., "African women and pesticides: more exposed to risks, less informed about the dangers". *Global Pesticide Campaigner*, Vol.4 No. 3, September 1994

Sudan

In two cases of poisonings, 87 people were poisoned in 1988 and 31 people died and 350 were poisoned in 1991 after eating bread made with organochlorine contaminated flour.

Pesticide containers are frequently reused for food storage. In 1988, 167 people were treated for acute organochlorine poisoning symptoms after eating cheese fermented in barrels that previously contained cotton pesticides; two people died from their symptoms.

Jamal, A., "Pesticide Tragedies in Sudan". *Global Pesticide Campaigner*, February 1994

Zimbabwe

The Zimbabwe Institute of Permaculture conducted a survey in 1997 of 30 poor smallholder farmers. All used an OP insecticide, dimethoate, on their crops which was applied without any protective clothing and all application techniques (56.7% used bare hands, 26.7% used knapsack sprayer and 100% used a brush, broom or tree branch and bucket) involved direct contact through handling, breathing and sometimes orally. Although it is illegal for pregnant women and children to apply the chemicals to the crops, they constitute the majority of persons applying the pesticides (Table 6).

Table 6. Pesticide users.

Family member	%
Women	80.0
Pregnant women	60.0
Children (under 10 years)	10.0
Men (over 16 years)	36.7

1. Storage
Most pesticides are stored insecurely. Children had access in 51% of all cases; 43.3% of the households studied store their pesticides in the garden; 30% store them in the bedroom; 16.7% in the kitchen; 13.3% in the storeroom; and 16.7% in the granary.

2. Poisonings
The study found 12 deaths involving accidental and intentional (suicidal) drinking, inhaling and dermal contact with pesticides. All farmers interviewed (30) had at some time or another experienced pesticide poisoning without receiving professional medical advice. The symptoms experienced after spraying included headaches, sneezing, vomiting, nausea, skin irritations and breathing problems.

One farmer is reported to have died after vomiting profusely. He had been drinking beer while spraying his crops. Two farmers were reported to have died after mixing pesticides with water and stirring the mixture with their bare hands. That these deaths were due to poisoning was confirmed by autopsy reports.

One farmer experienced severe burns when his knapsack burst, drenching his back in chemicals. Another developed an asthma-like condition after repeatedly winnowing grain treated with a grain protectant (Cooper-Shumba). She had not suffered from this condition before she was married, her parents' village had never used pesticides.

Zimbabwe hazards — too close to home, Zimbabwe Institute of Permaculture, *Pesticides News* 37, September 1997, Page 3

Europe

The European Federation of Agricultural Workers (EFA) represent the interests of two million agricultural workers throughout the European Union. In 1996 it conducted a survey among its members and over 1230 responses from individuals were analysed (Table 7). The conclusion was that the survey clearly shows the continuing hazards faced by those who work with pesticides on a daily basis, and the need for better training, increased health surveillance and above all, safer products and more non-chemical alternatives.

Ruelle, P., Health and safety concerns from European survey of operators, *Pesticides News* 36, June 1997, page 7

Table 7. Symptoms most often reported by pesticide users in a survey by the EFA. (Other symptoms occurred in 10% of cases: notable symptoms linked with the nervous system such as fatigue, difficulty in concentration, difficulty in muscle control and coordination of movement, trembling; and the respiratory system (allergies, different respiratory problems). As there are often multiple symptoms, the overall total of percentages exceeds 100%.).

Symptom	%
Headaches	67
Skin irritation	39
Stomach pains	33
Vomiting	30
Eye irritation	25
Diarrhoea	15

North America

In 1996, 22 farm workers were hospitalised while harvesting grapes after being poisoned by drift during the application of pesticides in a nearby cotton field.

Another 225 workers were estimated to have been exposed to the pesticide mixture of chlorpyrifos, fenpropathrin and profenofos. The mixture of symptoms included vomiting and eye and nose irritation.

Pesticide Action Network North America Updates Service, September 1996

Pesticide Residues

The previous two sections have shown that the import and use of WHO Class Ia and Ib pesticides is frequently observed in developing countries. These highly hazardous chemicals can be applied by spray workers with inadequate training and equipment and in excessive quantities, resulting in unsafe levels of residues on the food products. Consumption of food with high levels of chemical pesticide residues can cause acute poisonings while the chronic effects of consuming lower levels of pesticides over a long period of time are still not fully known. In addition, local economies can suffer when exports from developing countries are adversely affected through the enforcement of embargoes following the discovery of high residues.

The existence of pesticide residues on foodstuffs is a significant problem already recognised in developed countries in that many have established programmes which undertake regular testing of food samples for residues. Most developing countries are unable to set up programmes to regularly monitor pesticide residues on food. However, the size of the problem is indicated by one-off studies conducted in some developing countries and by data from the monitoring of residues in food imports by agencies such as the Food and Drug Administration in the US.

The following cases give results of residue analyses which have found unsafe levels of pesticide chemicals in a significant proportion of food samples. There is an evident risk of consumer exposure and cases exist of acute pesticide poisoning caused by pesticide residues on food. There is also the problem that consumers can still be exposed to toxic pesticides banned in their own country through the import of residues on food products from another country where the pesticides are still used. The effects of pesticide residues are a concern for both developed and developing countries.

Latin America

Costa Rica

In a 1991 report by Rhône-Poulenc to the Environmental Protection Agency, dangerous residues of aldicarb, an acutely toxic pesticide, were found in bananas grown in Latin America, exceeding the US tolerance at levels up to 10 times the legal limit.

The use of aldicarb has a double impact on people, firstly via poisonings, animal kills or water poisoning and then secondly the residue scandal which can cause economic dislocation in the banana industry; past experience shows that embargoes, product recalls and declines in consumption caused by pesticide residue problems often have drastic effects on communities.

Seidenburg, K., High levels of Aldicarb found in bananas, *Global Pesticide Campaigner*, June 1991, Vol. 1, No. 3

In 1993, Costa Rica's Plant Protection Service found that pesticide residues were found in 55% of food samples and 11% of the samples exceed maximum residue limits (MRL).

Between 1985 and 1991, more than 500 US tons of agricultural crops were detained in US ports because they surpassed the FDA's MRL.

Agne, S. (1996) *Economic Analysis of Crop Protection Policy in Costa Rica.* Pesticide Policy Project Publication Series No. 4, September 1996, pp. 31–32. University of Hannover

Asia

China

In 1995, OP contaminated vegetables were responsible for 200 students in Guangxi requiring hospital treatment.

Grimes, A., *Crop Production Opportunities in China*, Report DS147, Agrow, PJB Publications *Pesticides News* 39, March 1998, page 14

In 1987, the Health Department found that an outbreak of 120 cases of acute organophosphate poisoning in Hong Kong was caused by residues of metamidophos residues on vegetables imported from China. Since 1987, another 600 poisonings have been attributed to the same cause.

Pesticide Action Network North America Updates Service, May 19, 1995

Japan

Despite being banned, researchers from Kyushu University found that residues of organochlorine insecticides are still appearing in human breast milk in Japan.

This may be due to the persistence of organochlorines in the environment, residues from imported food or accumulation after being carried by air currents from other countries.

Agrow, *World Crop Protection News*, September 18 1998, page 18

India

Hundreds of people die from pesticide poisoning each year. A survey of pesticide residues in food samples collected in 12 states found residues in 85% of samples with 43% above the recommended doses.

Saloke, V.M. (ed.), *Safe and Efficient Application of Agro-chemicals and Bio Products in South and South-East Asia*, Proceedings of the International Workshop, 28-30 May 1997, Asian Institute of Technology, Thailand.
Pesticides News 38, December 1997, page 16

A seven year study by the Indian Council of Medical Research released in 1993 analysed 2205 cow and buffalo milk samples from 12 states. HCH (lindane) was detected in about 85% of the samples, with up to 41% of the samples exceeding tolerance limits. DDT residues were detected in 82% of the samples and 37% contained residues above the limit of 0.05 mg/kg, in some cases 44 times higher at 2.2 mg/kg.

"Facing a Silent Spring", *Global Pesticide Campaigner* 1997, Vol. 7 No. 2

Pakistan

Table 8. Comparative assessment of pesticide residue level in fruit and vegetables in Pakistan.

Area/Province	Number of samples	Number contaminated	Number above MRL
Karachi (Sindh)	250	93	45
North West Frontier Province	154	54	22
Islamabad	96	48	11
Quetta/Pishin (Baluchistan)	50	19	1

A four year testing programme on Pakistani fruits and vegetables showed maximum residue limits were regularly exceeded. Of 550 samples analysed for OPs, OCs and pyrethroid insecticide residues, 214 contained residues of which 79 (14%) exceeded the maximum residue limit (MRL) set by FAO, which could in some circumstances pose a hazard to the consumer (Table 8). Exceeding the MRL indicates good agricultural practice has not been carried out. Most significant overuse involved synthetic pyrethroids, with MRL exceeded by 20 to 30 times in some vegetables.

Pesticide residues in foodstuffs in Pakistan: Organochlorine, organophosphorus and pyrethroid insecticides in fruit and vegetables, Richardson, M. (ed.), *Environmental Toxicology Assessment*, Taylor and Francis, 1995, 438 pp.
Pesticides News 29, September 1995, page 19

Thailand

Between 1982 and 1985, government studies to monitor pesticide residues in food found that out of 663 samples tested covering nine food groups, 52% contained pesticides. DDT was detected in all nine groups (38%) and dieldrin was found in 15% of all food samples. Another survey by the National Environment Board in 1998 found pesticide residues in samples of soil, water, fruit, vegetable and field crops (Table 9).

Table 9. Pesticide residues in plant products and the environment in Thailand, 1998.

Sample type	Number analysed	Percentage containing pesticide residues
Soil	76	100
Water	139	86
Fruit	34	32
Vegetables	246	25
Field crops	71	17

On the basis of loss of produce which contains residues in excess of the MRL and therefore cannot be marketed and the costs of monitoring and residue control, Jungbluth (1996) estimates the costs of pesticide residues in food alone at 5065 million baht (US$202.6 million). Costs associated with pesticide residues in the environment are more difficult to estimate as there is insufficient information on the severity of pollution and its long-term effects.

Jungbluth, F. (1996) *Crop Protection Policy in Thailand. Economic and Political Factors Influencing Pesticide Use*. Pesticide Policy Project Publication Series No. 5, December 1996, pp. 29-33. University of Hannover

Africa

Chad

Four people died and six others were hospitalised with poisoning symptoms from suspected pesticide residues on salad and leaves used in a family meal.

Pesticide poisoning kills four in Chad, *Pesticides and Alternatives*, Pesticide Action Network Africa's Bulletin, 1999, No. 7

Egypt

Every year between 1986 and 1991, there have been fatal pesticide poisonings linked to the consumption of pesticide residues in food.

Pesticide Action Network North America Updates Service, May 19, 1995

Europe

Research findings announced by the UK Ministry of Agriculture, Fisheries and Food on 18 January 1995 have shown that unexpectedly high residues of OP insecticides occur in some carrots, exceeding MRLs in some cases.

Consumer Risk Assessment of Insecticide Residues in Carrots, Pesticides Safety Directorate, York, UK
Pesticides News 27, March 1995, page 3

North America

In 1985, 1000 US and Canadian consumers were poisoned by aldicarb residues in watermelon, the largest known incidence of pesticide food poisoning in North America.

Seidenburg, K., "High levels of Aldicarb found in bananas", *Global Pesticide Campaigner*, June 1991, Vol. 1 No. 3

An Environmental Working Group study utilised detailed government data on food consumption patterns and pesticide residues to conduct the first comprehensive analysis of the toxic dose that infants and children receive when the entire organophosphate family of insect killers is assessed in combinations, and at levels, that actually occur in the food supply. Based on the most recent government data available on children's eating patterns, pesticides in food, and the toxicity of organophosphate insecticides, the EWG estimated that:

- Every day, more than 1 million children age 5 and under (1 out of 20) eat an unsafe dose of organophosphate insecticides. Of these children, 100,000 exceed the EPA safe dose by a factor of ten or more.
- For infants six to twelve months of age, commercial baby food is the dominant source of unsafe levels of OP insecticides. OPs in baby food, apple juice, pears, apple sauce, and peaches expose about 77,000 infants each day to unsafe levels of OP insecticides.

While the amounts consumed rarely cause acut
"organophosphate" insecticides (OPs) have the potential to
damage to the brain and the nervous system, which are ra
extremely vulnerable to injury during fetal development,

childhood. This estimate very likely understates the number of children at risk because the analysis does not include residential and other exposures to these compounds, which can be substantial, and because EPA's estimates of a safe daily dose (the so-called references dose or RfD) are based on studies on adult animals or adult humans, and almost never include additional protections to shelter the young from the toxic effects of OPs.

The EWG analysis also identified foods that expose young children to the most toxic doses of these pesticides. It found that:

• One out of every four times a child age five or under eats a peach, he or she is exposed to an unsafe level of OP insecticides. Thirteen per cent of the apples, 7.5% of the pears and 5% of the grapes in the US food supply expose the average young child eating these fruits to unsafe levels of OP insecticides.
• A small but worrisome percentage of these fruits — 1.5–2% of the apples, grapes, and pears, and 15% of the peaches — are so contaminated with OPs that the average 25 pound one-year-old eating just two grapes, or three bites of an apple, pear, or peach (10 grams of each fruit) will exceed the EPA (adult) safe daily dose of OPs.
• Just over half of the children that eat an unsafe level of OPs each day (575,000 children) receive this unsafe dose from apple products alone.
• Many of these exposures exceed safe levels by wide margins. OPs on apples, peaches, grapes, pear baby food and pears cause 85,000 children each day to exceed the federal safety standard by a factor of ten or more.

Overexposed: Organophosphate Insecticides in Children's Food, 1998, Environmental Working Group

Using the FDA's own data, the EWG found that 5.6% of 14,923 samples tested by the FDA contained illegal pesticide levels. The FDA reported only 3.1%.

Over 90% of the violations reported in the EWG study involve two kinds of illegal pesticides: no-tolerance violations, where the pesticide is found on a crop even though the allowable level for the pesticide on that crop is zero; and over-tolerance violations, where the amount of the pesticide found exceeds the legal limit (or tolerance) for that crop.

Many pesticides that have been banned or restricted for health reasons were found illegally on scores of different foods. Examples include:

• Captan, a probable human carcinogen banned on 30 crops by the EPA for health reasons was found illegally on 14 crops.
• Chlorpyrifos (Dursban), a potent neurotoxin heavily used in schools and homes but restricted to use on certain foods to protect young children from additional exposure was found illegally on 16 crops.
• Endosulfan, a chemical cousin of DDT that mimics the female hormone oestrogen in the human body was found illegally on 10 crops.

US-grown produce is more than twice as contaminated with illegal pesticides the FDA reports. This two-fold underestimate is important because the lion's share of fruits and vegetables consumed in the US are domestically grown. For some major US crops the violation rates are well above average, including green onions at 16.7%, pears at 14.3%, tomatoes at 9.4%, green peas at 8.6%, and peaches at 6.1%.

The situation is similar for imported produce. The FDA reports 4.0% of imports with illegal pesticides; its records indicate that the rate is 7.4%. For countries which export tens of millions of pounds of produce to the US each year, the worst discrepancies include:

- Guatemala: an actual violation rate of 24.8% (due mostly to snowpeas and blackberries); the FDA reports 13.8%.
- Mexico: an actual violation rate exceeding 7.4%; the FDA reports only 4.0%.
- Canada: an actual violation rate of 5.0%; the FDA reports only 1.6%.

Environmental Working Group. Compiled from Food and Drug Administration Pesticide Monitoring Database FY 1992 and 1993. Surveillance data only.

Obsolete Pesticide Stocks

The UN Food and Agriculture Organization (FAO) has warned that huge amounts of obsolete and unused pesticides continue to threaten human health and the environment in many developing countries. FAO has urged the international community to increase its efforts to solve "this environmental tragedy".

FAO estimates that there are more than 100,000 tonnes of obsolete pesticides in developing countries, with 20,000 tonnes in Africa. The amount of stocks in the Near East countries is estimated at 5000 tonnes. Enormous stocks of pesticide waste also exist in Eastern Europe and parts of the former Soviet Union; several countries in Asia and Eastern Europe have stocks in excess of 5000 tonnes each. Due to the absence of environmentally sound disposal facilities stocks are constantly increasing. Some stocks are over 30 years old. In Africa particularly, large proportions of obsolete pesticides are left-over from earlier foreign assistance programmes. They can no longer be used because they are now banned or they have deteriorated as a result of prolonged storage.

Storage conditions rarely meet international standards. In many countries, pesticide containers are kept in the open, containers deteriorate and leak their contents into the soil, contaminating soil, water and groundwater. Most stocks are located in urban areas or near water bodies, putting ground water, irrigation and drinking water at risk. Many of these chemicals are so toxic that a few grams could poison thousands of people or contaminate a large area. Among the highly toxic and persistent pesticides in obsolete stocks identified by FAO were aldrin, DDT, dieldrin, endrin, HCH, lindane, malathion, parathion and others.

According to FAO, in Africa and the Near East only 1511 tonnes have been disposed of in 10 countries (Niger, Uganda, Madagascar, Mozambique, Zanzibar, Yemen, Tanzania, Zambia, Seychelles, Mauritania).

Total costs to remove obsolete pesticides from Africa alone are estimated at more than US$100 million. Most of the money spent on disposal of pesticides in Africa has been financed by the Netherlands, Germany and FAO. Denmark recently committed US$6 million for pesticide removal and capacity building. At present the agro-chemical industry contributions are very limited, but they are expected to grow in the near future.

The preferred way to dispose of obsolete pesticides is high temperature incineration. None of the developing countries — except for a few newly industrialised nations — have facilities for the safe and environmentally sound disposal of pesticides, and pesticides are re-packaged and shipped to a country with a hazardous waste destruction facility. In the past, waste has been shipped to Europe.

Unless prevention occurs, FAO warned, it is likely that accumulation of hazardous pesticides in the environment will continue unabated as the world-wide sales of pesticides increased substantially both in 1995 and 1996. Total global sales of pesticides for 1995 reached US$29 billion, of which pesticides worth nearly US$24 billion were sold by 11 major companies. Disposal costs vary from US$3500 to US$5000 per tonne.

According to FAO, the main causes for the accumulation of pesticides are:

* pesticides banned while in storage
* inability to forecast pest outbreaks and excessive donations
* poor assessment of pesticides requirements (donations made out of season)
* inadequate storage facilities and poor stock management
* ineffective or wrong pesticide formulations
* aggressive sales practices.

The long-term solution to disposal problems lies in preventing accumulation of obsolete pesticides, according to FAO. Stocks should be kept as small as possible and pesticide use should be drastically reduced. FAO has called upon its members to apply Integrated Pest Management (IPM) and to reduce the use of pesticides, where this is possible.

FAO Press releases: "FAO: Problem of obsolete pesticide stocks deserves greater attention by donor countries and industry" 98/15; "Dangerous pesticide stocks removed from Zambia and the Seychelles — large stocks continue to threaten health and environment, FAO says" 97/31; and "FAO says huge stocks of obsolete pesticides threaten environment and public health in developing countries" FAO/3634, 5 June 1996

The FAO estimate that initial clean-up costs of obsolete pesticides will be at least US$100 million in Africa alone does not appear to take into account the costs of complete site detoxification, which could be considerable. In Sudan for example, an estimated 500 kg of soil needs to be detoxified for each kilogram of obsolete pesticides. Unfortunately, developing country governments do not have the financial resources for dealing with this problem themselves.

"Obsolete Pesticides", *Global Pesticide Campaigner*, Vol.2 No.1

Africa

A sizable proportion of the stocks listed in the table below contain persistent organochlorines, such as DDT, dieldrin and lindane, which are banned in many countries (Table 10). Obsolete stocks also contain acutely toxic organophosphates such as fenitrothion and malathion, as well as many chemicals with no identification whatsoever.

Originally published in *Pesticides News*, December 1991 and updated in *Global Pesticide Campaigner*, Vol 2 No1 in February 1992

Table 10. FAO *preliminary* inventory of obsolete pesticide stocks requiring disposal.

Country	Total (metric tonnes)	Total includes the following known pesticides	
Algeria	937	900	HCH*
		37	carbaryl
Benin		2,500 litres	dieldrin*
Botswana	31	18	fenitrothion
Burkina-Faso	93	40	TMTD and heptachlor*
Cameroon	240	91	fenitrothion
Cap-Vert	21	15	trichlorfon
Chad	101	55	lindane
		39	HCH*
Ethiopia	295	147	unknown
Gambia	10	5	unknown
Ghana	36	5	parathion*
Guinea-Bissau	33	10	fenitrothion
Ivory Coast		2,500 litres	dieldrin*
Kenya	48		
Liberia	nil		
Libya	>300	300	HCH*
Madagascar		48,000 litres	dieldrin*
Malawi	125	70	DDT*
Mali	116	82	dieldrin*
Mauritania	242	204	dieldrin*
Morocco	>1862	1862	HCH*
Niger	31,060 litres		
Senegal	155	55	HCH*
		46	dieldrin*
Somalia	103		
Sudan	980		
Tanzania	140	77	malathion
Tunisia	>500	500	HCH*
Uganda		51	dieldrin*
Yemen	207	77	dieldrin*
Zambia	>85	10	DDT*

* On the FAO's prior informed consent (PIC) list.

Ethiopia

In May 1992, the total pesticide stocks (current and obsolete) in Ethiopia amounted to nearly 5400 tonnes including 652 tonnes of "Dirty Dozen" pesticides, mainly DDT, lindane, heptachlor, aldrin, chlordane and paraquat.

"Demise of the Dirty Dozen", *Global Pesticide Campaigner*, 1995, Vol.5 No.3

Somalia

In 1978, Somalia received an unsolicited donation of pesticides which it never used. The pesticides are no longer usable and now require expensive disposal.

"Obsolete Pesticides", *Global Pesticide Campaigner*, Vol.2 No.1

Sudan

A pesticide store in Hassahisa consists of two open sheds with a dirt floor and is not designed for prolonged storage. Approximately 500 tonnes of obsolete chemical pesticides are now lying in rusting and breaking containers. Hassahisa is a densely populated area only 1.5 km from the Blue Nile and less than 500 m from the Gezira main canal. Residents have complained since the mid-1980s of the strong smell and a Sudan Development Association survey of nearby residents found that 95% have experienced headaches, nausea, dizziness, dazzling, increases in allergies and loss of appetite. In addition, over 60% of the families interviewed had lost domestic animals and poultry from drinking contaminated water.

Sudan Development Association report to Oxfam, December 1993, in *Pesticides News* 24, June 1994
"Obsolete Pesticides", *Global Pesticide Campaigner*, Vol.2 No.1

Tanzania

Obsolete stocks of pesticides are estimated at more than 90 tonnes, including approximately 60 tonnes of DDT. The country lacks appropriate disposal facilities and thus obsolete pesticides have either been dumped indiscriminately or stored indefinitely.

"Dumped pesticides persist in Tanzania" *Pesticides News* 37, September 1997

Asia

Nepal

Large quantities of banned and expired pesticides have been sent to Nepal as "aid" from both developed and developing countries. For example, Indonesia exported 100 tonnes of old stocks of DDT to Nepal in 1994, despite protests from concerned environmental and public interest groups in both countries.[1] In addition to DDT, pesticides "given" to Nepal as aid include other organochlorines such as BHC, as well as the organophosphates malathion and parathion. As a result, Nepal has been forced to take on the burden of disposing of other countries' unwanted pesticides, as well as its own unmanageable quantity of old stocks.

Since 1962, agrochemicals and chemical fertilisers have been distributed in Nepal by the Agriculture Inputs Corporation (AIC), a public enterprise owned by the government. Private sector traders also play a major role in pesticide distribution, reselling agrochemicals purchased from AIC. In 1994, AIC distributed about 335 metric tonnes of dust and granulated pesticides and about 2460 litres of liquid pesticides.[2]

As a result of years of accumulation, AIC has large quantities of expired pesticides stored in various government-owned warehouses throughout the country. In many cases, pesticides are stored in inadequate facilities, including warehouses located in populated areas adjacent to dwellings, schools and livestock. Some of these warehouses are in serious need of repair with crumbling walls and leaking roofs, and are especially vulnerable to natural disasters such as floods, fires or earthquakes. Since these stocks of old chemicals present a serious risk both to human health and the environment, in 1988 the UN Development Programme provided US$577,000 to the government of Nepal for technical assistance to dispose of stocks of old pesticides.[3] The funds, administered by the Manila-based Asian Development Bank (ADB/Manila), were used to hire the New Zealand firm, ANZDEC Ltd., to develop a disposal plan for the chemicals. ANZDEC found approximately 175 metric tonnes of expired pesticides stored in the country,[4] and prepared a plan made up of several disposal methods including burning in open fires, burial, spreading, re-processing and incineration.

The entire process, from acceptance of pesticides through storage and disposal, lacked any co-ordination and consultation between government agencies, technical experts, environmentalists and farmers. Pesticides that were unneeded and unwanted by farmers have been purchased or accepted as aid; expired stocks were allowed to enter the country. Rarely were environmental or health impacts considered at any stage in the process.

[1] *The Pesticide Trail, The Pesticides Trust, 1995.*
[2] *AIC. Personal communication, March 1995.*
[3] *NEFEF-Pesticides Watch, 1993. Playing with Poison.*
[4] *AIC. Personal communication, March 1995.*

Summarised from "More Deadly Donations: Disposal of Expired Pesticides in Nepal" Ananda Tamrakar. *Global Pesticide Campaigner*, Volume 5, Number 2, June 1995.

Index

A

Aldicarb, 18, 19, 38, 41
Aldrin, 22, 45, 48
Argentina
    Pesticide sales, 6
    Pesticide use, 6
Atrazine, 16

B

Bacillus thuringiensis, 10
Bananas, 8, 18, 19, 38
Banned pesticides, 3, 5, 15, 17, 21, 32,
    37, 38, 42, 45, 46, 47, 48
Benzene hexachloride, 22
BHC, 11, 49
Brazil
    Pesticide market structure, 7, 10
    Pesticide sales, 7
    Pesticide use, 7

C

Captan, 18, 42
Carbamates, 12, 19, 29
Carbofuran, 22
Chile
    Pesticide imports, 7

China
    Pesticide sales, 10
    Pesticide use, 10
Chlordane, 48
Chlorpyrifos, 35, 42
Cholinesterase inhibition, 17, 31
Coffee, 7, 13, 17, 31
Colombia
    Pesticide production, 7
    Pesticide sales, 7
Costa Rica
    Imports, 9
    Pesticide imports, 8
    Pesticide market structure, 9
    Pesticide sales, 8
Cotton, 3, 7, 10, 12, 14, 23, 31, 33, 34

D

DBCP, 20
    Male sterilisation, 19
DDT, 5, 10, 11, 32, 39, 40, 42, 45, 47,
    48
Diazinon, 14
Dichlorvos, 16, 18
Dieldrin, 23, 40, 45, 47
Dimethoate, 26, 33
Dirty Dozen, 11, 20, 23, 24, 26, 48

E

Ecuador
    Pesticide sales, 9
Endosulfan, 11, 17, 32, 42
Endrin, 22, 23, 45
Exporters
    Colombia, 9
    Germany, 9
    Japan, 6
    Switzerland, 9
    UK, 6
    USA, 5, 9

F

Fenitrothion, 14, 47
Fenpropathrin, 35
Flower industry
    Colombia, 17
Food contamination, 23, 24

G

Glyphosate, 5

H

HCB, 10
HCH, 11, 39, 45, 47
Health costs, 28
Heptachlor, 22, 47, 48
Herbicide tolerant crops, 5, 6

I

Importers, regional, 6
Imports
    Chile, 7
    Costa Rica, 8, 9
    Kenya, 13
    Madagascar, 14
    Pakistan, 11
    Thailand, 12
India
    Pesticide market structure, 11
    Pesticide production, 10, 11
    Pesticide sales, 11

J

Japanese aid, 6

K

Kenya
    Pesticide imports, 13
    Pesticide sales, 13

L

Lambda-cyhalothrin, 14, 18
Lindane, 22, 24, 32, 39, 45, 47, 48
Locusts, 14

M

Madagascar
    Imports, 14
Malathion, 11, 45, 47, 49
Mancozeb, 11, 18
Metamidophos, 22, 24, 26, 38
Methyl parathion, 11, 27, 30, 32, 45,
    47, 49
Minimum tillage, 5, 7
Mirex, 5
Monocrotophos, 11, 23

N

Naled, 18

O

Occupational pesticide exposure. *See*
    Pesticide exposure, occupational
Organochlorines, 10, 12, 22, 33, 38, 39,
    47, 49
Organophosphates, 12, 17, 19, 22, 23,
    24, 26, 29, 31, 33, 38, 39, 41, 42, 47,
    49

P

Pakistan
    Imports, 11
    Pesticide market structure, 11, 12
    Pesticide sales, 11, 12
    Pesticide use, 11, 12
Paraquat, 25, 26, 48
Pentachlorophenol, 10
Peru
    Imports, 5
    Pesticide sales, 9
Pesticide adulteration, 12, 14

Pesticide donations, 6, 14, 48
Pesticide exposure
  Fatalities, 15, 17, 19, 21, 23, 24, 25,
    26, 29, 32, 33, 34
  Health effects, 16, 17, 18, 19, 25, 26,
    27, 30, 31, 33, 34, 35
    Children, 16, 17, 18, 19, 22, 23
  Occupational, 16, 17, 18, 23, 24, 26,
    27, 28, 29, 31, 32, 33, 34
    Women, 17, 18
  Routes, 20
  Unsafe practices, 24, 26, 27, 31, 32,
    33, 34
Pesticide market structure
  Brazil, 7, 10
  Costa Rica, 9
  India, 11
  Pakistan, 11, 12
  Thailand, 12
  Vietnam, 13
Pesticide poisonings
  Brazil, 16
  China, 23
  Colombia, 17
  Costa Rica, 19, 20
  Guatemala, 21
  India, 24
  Indonesia, 25
  Latin America, 16
  Malaysia, 25
  Mexico, 21
  Middle East, 23
  Nicaragua, 22
  North America, 34
  Pakistan, 26
  Philippines, 27
  Residues in food, 23, 24, 38, 41
  Thailand, 28
  Under reporting, 16, 19, 22, 23, 24,
    25, 28
Pesticide poisonings, global, 2, 15
Pesticide production
  Colombia, 7
  India, 10, 11

Pesticide residues, 39
  Bananas, 38
  Detention of crops, 38
  Economic effects, 40
  Economic effects, 38
  Imported food, 43
  Levels in food, 40, 41, 42, 43
  Levels in food, 38, 39
  Poisonings, 38, 41
Pesticide resistance, 8, 10
Pesticide sales
  Argentina, 6
  Brazil, 7
  China, 10
  Colombia, 7
  Costa Rica, 8
  Ecuador, 9
  India, 11
  Kenya, 13
  Pakistan, 11, 12
  Peru, 9
  Senegal, 14
  South Korea, 12
  Thailand, 12
  UK, 6
  Vietnam, 13
Pesticide sales, global, 2, 4
Pesticide use
  China, 10
  Pakistan, 11, 12
  South Korea, 12
Pesticide use, global, 4
Pesticides, obsolete, 2, 45
  Disposal, 45, 49
  Disposal costs, 46
  Health effects, 48
  Health risks, 45
  Pesticide donations, 48
Phorate, 11
Profenofos, 35
Pyrethroids, 10, 12, 22, 39, 40

R

Rice, 3, 14, 27, 30
Roundup. *See* Glyphosate

S

Senegal
  Pesticide sales, 14
South Korea
  Pesticide sales, 12
  Pesticide use, 12
Sterilisation, 19
Sugar cane, 14
Suicide, 16, 32, 33

T

Thailand
  Imports, 12
  Pesticide market structure, 12
  Pesticide sales, 12
Tobacco, 3, 14, 26

V

Vietnam
  Pesticide market structure, 13
  Pesticide sales, 13

W

Wilms' tumours, 16
World Health Organization, 2, 3, 15,
    16, 17, 22, 27
  Pesticide classifications, 3, 5, 9, 12,
    14, 17, 26, 27, 30, 31